Animals in Australia

by Louise Crary

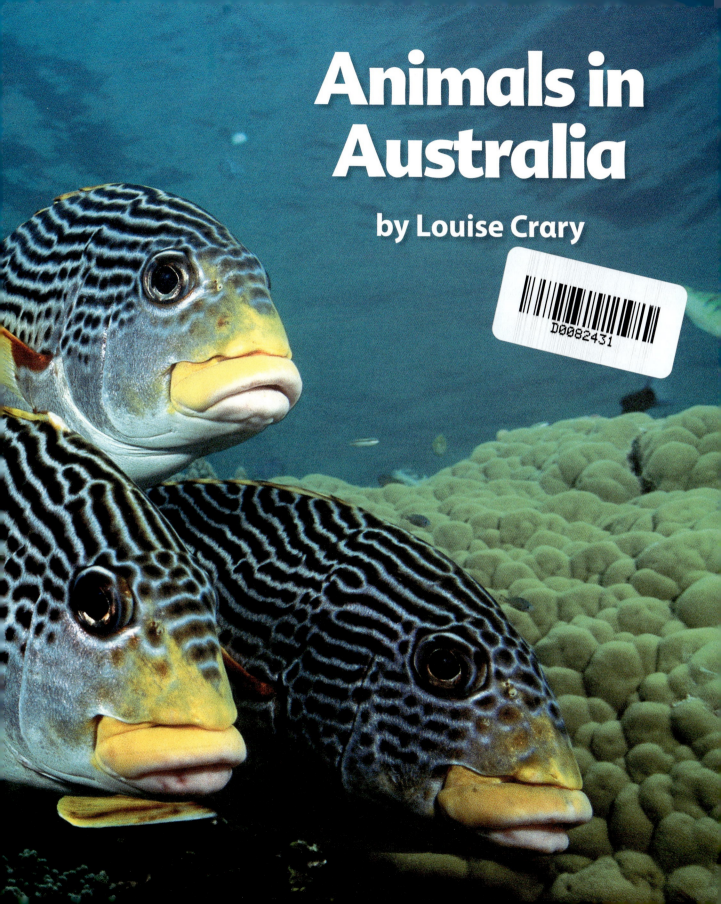

Many animals live in Australia.

Crocodile

Lorikeet

Dingo

Stingray

3

Kangaroos live in Australia.

Tails help kangaroos stand.

tail

Emus live in Australia.

Legs help emus run.

Fish live in Australia.

Fins help fish swim.

fin

Koalas live in Australia.

Claws help koalas climb.

claw

Birds live in Australia.

wing

Wings help birds fly.

Sea lions live in Australia.

Fur helps sea lions stay warm.

fur

Animals in Australia

tail

claws

legs

wings

fins

fur